SUSTAINABLE FUTURES

SUSTAINABLE FUTURES

Technological Solutions for a Green Planet

B. VINCENT

QuantumQuill Press

CONTENTS

CHAPTER 1

Introduction

The purpose of this report is to explain where technological advances are taking us to make a more sustainable future. This means a future where the inhabitants of the planet would be able to sustain their way of life without depleting the natural resources required to do so. This is a major question as, in today's society, we are finding it harder to do so as the required standard of life is becoming higher for many. High-powered and energy-efficient hand dryers to air dry hands in public toilets, rather than using paper towels, are a simple yet effective example of what a sustainable way of life would be. This has been achieved by a university project known as "Integrating Innovation for Sustainability," run between Loughborough, Surrey, and Cranfield. This is just the very start of where we could be heading, and there are many great ideas that could take us to the target of a more sustainable way of life.

Talk of global warming and the need for more sustainable practices is everywhere - on the news, in schools, and in politics. It is hard to escape the topic these days, and it should not be ignored. With all this talk, many people are left to wonder what it means to be "green" and, more importantly, what they can do. For many, the first thing that comes to mind is solar panels, wind turbines, or electric cars - all of which are associated with technological advances.

CHAPTER 2

Renewable Energy Sources

The power available from a hydroelectric plant depends on two factors: the rate at which water flows and the amount of change in height the water undergoes. A large increase in height and flow rate results in a large amount of available power. High head hydro plants require a smaller amount of water flow and hence a reduced environmental impact. The water can be dammed, generating a great height on one side and then be released into the other side. The stored water can be used as a store of energy and released when required. This allows the rate of electricity generation to be regulated to cope with changes in demand. This is a great advantage over other forms of renewable energy. Hydroelectric power has no fuel costs or adverse effects on the environment and does not use limited natural resources.

The energy available from moving water can be harnessed and converted into electricity. This is the principle behind hydroelectric power. Water is the primary renewable energy source in the world. Hydro power has been in use for thousands of years. Its current popularity is due to the long-term costs and the limited adverse effects on the environment. Today, hydroelectric power is the most important form of power generation from water. It produces over 15% of the world's electricity and more than 65% of all renewable electricity. As far as producing greenhouse gases and pollutants, it is close to the perfect energy source. It is widely regarded as a proven, reliable, and cheap source of electricity. Hydro plants can be easily regulated to cope with changes in demand and are used as base load energy suppliers. This means its share in the world's power supply may be increased if other forms of energy are reduced.

Wind energy is the world's fastest-growing energy source. It is both renewable and clean and produces no air or water pollution. This energy is extracted from airflow using wind turbines or sails to produce mechanical or electrical energy. Windmills are used for their mechanical power, water pumping or drainage, or sails to propel ships. The total amount of economically extractable power available from the wind is considerably more than present human power use from all sources.

Solar power is energy that is derived from the sun and converted into sunlight. It is an enormous and inexhaustible resource. Solar power generation has several advantages over other forms of electricity generation. It does not pollute the air. Conventional electricity generation can cause air pollution that degrades the quality of the air and water. Global electricity production is the single largest source of atmospheric pollution. Solar power generation does not require water, and this is important since the availability of clean water is one of the Earth's most pressing issues. Today, nearly one-third of the world's population is living in areas with water scarcity.

2.1 Solar Power

Moving to more revolutionary methods of solar energy generation, a Sydney-based company called EnviroMission has recently transplanted and begun testing a new solar technology called the Solar Tower. The tower is essentially a large greenhouse in which the air is heated by the sun, making it rise and pass through turbines located at the base, thus generating electricity. In agreement with a 2009 press release, EnviroMission has signed a deal with the Arizona Strip, an area in the USA near the Grand Canyon, to build a large-scale version of the Solar Tower. This technology is hoped to eventually become a cost-effective method of large-scale power generation and has the potential to be one of the most efficient solar power systems.

A common technique used to exploit solar power is the installation of solar panels, which are designed to collect the energy emitted by the sun. The concept is simple, but the efficiency is not. Solar energy technologies come in many forms including heating water for domestic use, heating spaces, and generating electricity. The growth of solar generated electricity is of particular interest, especially after the announcement of the Australian Government's (at the time of writing) Solar Credits Program. This scheme provides a tradable certificate to those who install solar powered systems and is an attempt to double the amount of solar power generated in Australia by 2010. These certificates reduce the initial cost of system installation and effectively increase the price of standard electricity, making it more economically viable to generate solar power. In addition, this is an implementation of the polluter pays principle, with proceeds of the carbon emissions trading system to be used to fund the program.

2.2 Wind Power

Despite being a renewable technology with few environmental and sustainability-related drawbacks, the key limitation of wind power is its inherent variability and unpredictability. On a small scale, wind power can be highly variable, and therefore wind farms are often coupled with battery or other energy storage systems. On large electricity grids, though, this output variability at the individual generator level is less of an issue, but wind energy could represent a greater proportion of electricity output and thus have complex effects on supply and demand stability. This is something that will be addressed with an electrical smart grid and expanded infrastructure in demand management, storage, and interconnection, and will likely lead to increased penetration of wind power in years to come.

Designs of wind turbines and wind farms vary, but they can be divided into two main types: either onshore or offshore. Onshore wind farms are generally cheaper to install, have a less complicated installation process, making them ideal for developed countries, and produce moderate energy outputs. However, offshore wind farms are able to capture more energy due to the higher average wind speeds over the ocean, and wind speed's strong correlation with energy output is a major advantage to offshore farms. Also, they don't take up any land area that could be used for other purposes. As a result, the development of this technology is strong, and it is increasingly being seen as a major potential energy source for many countries around the globe.

Wind power is an indirect form of solar power – the sun's radiation heats different parts of the earth at different rates, leading to the development of wind currents. The kinetic energy of the wind can be harnessed into mechanical energy, which can drive a variety of machinery. Early forms of wind technology allowed for the mechanization of many farming processes, such as milling, and the transportation of goods via boats. The development of the wind turbine, which is a device that converts the kinetic energy of the wind into mechanical energy, has led to the possibility of using wind power as a form of electricity generation. Modern wind turbines connected to electrical power generators can produce large amounts of electricity. These 'wind farms' now represent a mature and cost-effective technology that is the fastest-growing form of renewable electricity generation in the world.

2.3 Hydroelectric Power

The cost-effectiveness of large-scale hydroelectric power is increasing due to constantly depleting resources and the ever-increasing costs of fossil fuels.

Hydroelectric dams are usually constructed on rivers and are built to store water in a reservoir. These reservoirs are quite often used as multi-purpose lakes, and are used for food, water supply, irrigation, and recreation, as well as electricity generation.

The most common methods of hydroelectric power generation are using a dam to retain a large amount of water in a reservoir, and then letting it out through turbines. Another way is diverting a portion of the river's water through a channel or a series of pipes to a lower point, and then through a turbine before returning it to the river.

Hydroelectric power is one of the most viable and sustainable sources of energy available to us. With the power from the flowing or falling water, you can generate power. This power generation can take place in many forms, ranging from the small water wheel to power a single home, to using a screw and a small generator with run-of-river, up to a large-scale hydroelectric dam which will provide large-scale electrical power.

Also, the summary reflects that hydroelectric power converts the gravitational potential energy of falling water into electricity, using gravity to cause water to fall or to continuously ease the river. Additionally, hydroelectric power converts energy from turbines into mechanical or electrical energy, giving the possibility of using an animal to provide mechanical power.

CHAPTER 3

Energy Storage Technologies

CAES involves converting electrical energy into compressed air, which is then stored in underground caverns. The air can later be released and heated before passing through an expansion turbine to generate electrical energy. While still in the testing stages, various theoretical CAES systems can have very high efficiencies and low costs, making this a very promising technology for the future.

Being able to store energy is a big issue for many governments and researchers to this day. The main issue with energy storage lies in if the amount of energy coming out of the system is less than the amount of energy going in. Currently, the only form of electrical energy storage that is relevant today is pumped hydro storage. This is where during times of low energy demand, water is pumped up to a reservoir and then let out during time of high demand to turn turbines and thus generate electricity. This has very high efficiency and is the current main form of grid energy storage in the world. On a smaller scale, batteries are commonly used. However, the forms of which energy is stored, such as lead acid batteries, have a very low round trip efficiency and can cause pollution if not disposed of responsibly. This, while helping in small scale energy storage, is not useful in the greater picture of sustainability.

A step up from these methods is the current research into Compressed Air Energy Storage (CAES) and future batteries.

3.1 Battery Storage

Sodium-sulfur batteries are the top battery technology for stationary energy storage. These batteries are widely used in Japan and are the best energy storage technology for large-scale renewable energy plants. Sodium-sulfur batteries offer high round trip energy efficiency, which is a measure of how much energy is returned for every unit of energy used. High round trip energy efficiency essentially means less energy wastage and more cost savings. Sodium-sulfur batteries are limited in the fact they must be operated at very high temperatures and cannot supply or store power at rapid rates. However, continuous battery development should mean sodium-sulfur batteries are a key contributor to the energy storage market in the future.

Lithium-ion batteries are well established and regarded as the best technology for the growing storage market. Lithium-ion batteries provide high specific energy (energy per unit weight) and energy density (energy per unit volume) at a good cost. This makes them the leading technology for portable devices such as smartphones and laptops. The inherent ability for lithium-ion batteries to be scaled up and provide high capacity electrical storage for a cost-effective price makes them the ideal candidate for storing renewable energy. All forms of lithium-ion batteries are still an expensive technology. Prices have fallen hugely over the years and are expected to continue to fall. High investment and rapid growth in the electric vehicle industry should mean that lithium-ion battery prices continue to decrease over the coming years. This is highly significant to the renewable sector because it means lithium-ion batteries will be providing a low-cost, high-capacity energy storage solution.

3.2 Pumped Hydro Storage

Pumped hydro storage consumes electrical power to pump water from a lower reservoir to an upper reservoir for storage. Upon discharge, the water is released back down to the lower reservoir using a hydro turbine which generates electricity. Data provided by the International Energy Agency shows that pumped storage capacity in Europe is expected to increase by approximately 60 GW by the year 2050. This represents the single largest expected increase in installed storage capacity of any storage technology. The increasing role of renewables in Europe is the primary driving force behind this investment in pumped storage. In the future, energy storage will become essential to enable the managed use of renewable generation to provide supply security and reliability. Pumped storage development is expected to play an increasingly important role in helping to achieve this by ensuring that energy produced from renewable resources is fully utilized.

Pumped hydro is the most important and widely used storage technology globally, with a share of about 99%. It is well established and has a good record in the areas of system reliability, response time, and ancillary services. Pumped hydro storage is applicable over a wide range of storage capacities from a few MWh to several GWh, and it can provide power over an extended period, ranging from a few hours to several days. This technology will also become more important in the future as it has the advantage over other forms of storage in that it can contribute to peak generation capacity. With the current strong emphasis on renewable energy technologies, there will inevitably be an increasing requirement to manage renewable electricity generation in a way that fully harnesses the available energy resource. This will involve a shift from simply load following to storing surplus energy during periods of high generation and using it to provide power at times when generation is insufficient.

3.3 Compressed Air Energy Storage

Future generation facilities and storage concepts will gain insight on CAES. Advanced and service-driven power cycles, such as combined cycles and integrated gasification combined cycles, with high levels of process heat which contribute to the production of electricity, will seek methods of energy storage. An aging and phasing out electrical infrastructure will seek ways to level peak power demands and transition to a more economical and stable future electricity system. With its large energy storage potential and environmentally friendly non-fuel burning nature, CAES will make a significant impact in the future global energy market.

Tests and simulations have shown that CAES has great potential for energy storage and load following when using the diabatic approach. One study in particular simulated the operation of CAES with a wind farm in which a high portion of wind energy was stored. The high ramp rates and high cycle efficiencies associated with the diabatic approach made CAES an ideal candidate to smooth the wind power variations. Simulation results concluded that adiabatic efficiencies as high as 80% were attainable. This study also identified various occurrences that resulted in wasted energy in the adiabatic process. Implementation of a new heat exchanger design and materials of construction with advanced high-temperature capabilities consumed less of the stored energy. As well, machinery and component design optimization resulted in fewer pressure drop occurrences in the storage cycle. Furthermore, expansion turbine design optimization increased the turbine inlet temperature, which also reduced the specific energy consumption. All of these findings pointed to further research and development of an adiabatic CAES cycle. Direct studies comparing this adiabatic CAES cycle in compression and expansion to the current conventional approach are currently underway to determine its overall technical and economic impact.

Compressed air energy storage (CAES) is a mature technology with two current methods of execution. One development approach is to compress the air in an underground cavern. The other direct and indirect approaches have air compressed on the surface, with the former storing the air in tanks and the latter employing a pressure vessel. When electricity is needed, the stored compressed air is released and heated. The heated air expands in an expansion turbine that drives an electrical generator. The process is analogous to a simple gas turbine cycle. Often, recycled heat from the system is used, or natural gas is employed to supply the heat in the expansion process.

CHAPTER 4

Sustainable Transportation

The reduction in air pollution would be beneficial to human health as outdoor air pollution is a carcinogen that can also cause lung cancer. Particulate matter, nitrous oxides, and sulfur dioxide from the combustion of fossil fuels in internal combustion engine vehicles are the principal sources of urban air pollution. These can exacerbate asthma, cause acute respiratory diseases, and are contributing factors in the development of chronic respiratory diseases. In some instances, the health effects have more serious consequences, such as the onset of heart disease. By 2040, the global demand for energy is projected to increase by 37%. If this energy is supplied through the burning of fossil fuels, it will increase the emissions of air pollutants unless there is a substantial shift to renewable energy sources.

Electric vehicles have the potential to reduce CO_2 emissions, the most prevalent greenhouse gas causing human-induced climate change. Powering vehicles from electricity stored in batteries instead of gasoline would result in lower emissions in the production and electricity-based operation of vehicles. In examining the attributes of electric vehicles, it is helpful to comprehend the well-to-wheel emissions of greenhouse gases attributable to a vehicle's operation. These are the emissions produced throughout the fuel's production, distribution, and use. Studies comparing electric vehicle emissions to gasoline vehicles have resulted in a consensus that the emissions are, in fact, much lower for electric vehicles.

4.1 Electric Vehicles

The first practical electric vehicles were produced in the 1880s. Compared to internal combustion engine vehicles, electric cars are quieter, have no tailpipe emissions, and lower emissions overall. At the fuel economy equivalence level, advances in battery technology may see cost effectiveness in the future, while fuel cells only generate electricity at 50% efficiency. Since they are not reliant on internal combustion, EVs could be the only type of car when finite fossil fuel resources are exhausted. Electric trains and cars can draw power from a multitude of sources and if that primary source is renewable, the vehicle can be considered zero emission. In the long term, EVs are seen as truly sustainable. However, due to the present methods of electricity generation, the benefits are not always realized. Existing electric cars like the G-Wiz are not much more than motorized quadricycles and passenger cars face tough competition from fuel efficient hybrids due to the higher cost and limited range. However, increasing concern for energy security and the environmental effects of global warming may see interest and incentives in this technology rise. This may be further accelerated by new laws banning fossil fuel vehicles from certain metropolitan areas.

An electric vehicle (EV) is a vehicle which uses one or more electric motors for propulsion. Depending on the type of vehicle, motion may be provided by wheels or propellers driven by rotary motors, or in the case of tracked vehicles, by linear motors. Electric vehicles can include electric cars, electric trains, electric airplanes, electric boats, electric motorcycles and scooters, and electric spacecraft.

4.2 Hydrogen Fuel Cell Vehicles

A hydrogen fuel cell vehicle is an alternative fuel vehicle that uses a hydrogen fuel cell to power its on-board electric motor. Fuel cells in vehicles create electricity to power an electric motor, generally using oxygen from the air and compressed hydrogen. The benefits of hydrogen fuel cell vehicles include zero emissions, as they only produce water, less environmental impact, and reduced dependence on oil since hydrogen can be produced from a wide variety of resources. A hydrogen fuel cell uses a catalyst to separate electrons from a hydrogen atom. The electrons go through a circuit, generating electricity, and the ions go through the electrolyte to meet with the electrons and combine with oxygen from the air to form water. This process is similar to an internal combustion engine, but the combustion occurs at the anode with hydrogen and oxygen from the air, and there are no moving parts. This process has high efficiency, especially in comparison to internal combustion engines. The electricity is stored in a high-voltage battery to power the electric motor, and the only byproduct is water. This electricity can also be used to power electronic devices, and fuel cells are utilized in buildings as power generation units.

4.3 Biofuels

Biofuels are liquid fuels made from plants. They are near carbon neutral - the carbon released when they are burned is balanced by the carbon absorbed by the plants, so they do not contribute to global warming. The energy they give out is derived from the sun through photosynthesis, and so they are a sustainable source of energy. The use of biofuels would thus move us towards a solar-powered transport system. Currently, bioethanol is the main biofuel being used. This is an alcohol made by fermenting and distilling starch crops such as corn, barley, and wheat or sugar crops such as sugarcane and sugar beet. Biomethanol can be made through similar processes using organic matter. Both can be used in petrol engines to reduce emissions and in some places are available as petrol substitutes. However, the real promise for biofuels lies in biodiesel, which is made through transesterification of plant oils or animal fats. Biodiesel can be used in unmodified diesel engines and has superior fuel economy and lubricity while greatly reducing emissions, particularly particulate matter and hydrocarbons. It can also be used to run stationary diesel engines for electricity production. The vast majority of biofuels are currently not being produced sustainably, so there is a long way to go before they can offer a truly green alternative.

There are a number of reasons why we need to develop more sustainable modes of transportation. The transport sector is the fastest growing contributor of greenhouse gases, and this looks set to continue - by 2015, emissions from the sector are likely to have doubled since 1995. At the same time, our reliance on petroleum is ever increasing, and the supply of cheap oil is rapidly dwindling. This is likely to lead to increased geopolitical tensions as nations rely on ever more expensive oil imports. The inevitable upward pressure on oil prices will be most keenly felt by the world's poor, to whom motorized transport is often out of reach. Meanwhile, concerns over petroleum supply and price volatility increasingly trouble vehicle manufacturers. For these reasons, a sustainable energy source for the transport sector is now being sought.

CHAPTER 5

Smart Grid Systems

The final demand resource is to integrate distributed storage and generation mainly to ensure reliability; however, it can be cost-effective to provide services back to the system. An overall societal measure of the impact of demand response can be assessed by the resource value compared to the overall cost.

Demand response, which is the voluntary reduction of electricity consumption based on price signals, can be achieved by the deployment of smart appliances and equipment, enabling communication between the provider and the consumer indicating when electricity is at its highest cost and, in the case of renewable intermittent resources, when the supply is off-peak. This, in turn, allows energy services to be provided at a lower consumption of resources. This concept can be similarly replicated to track interruptible loads and energy efficiency programs. The ability to provide energy consumption feedback and information for a service can allow a decision of whether to run it at times of high energy cost or find an alternative, e.g., using solar energy. This method is highly transparent, and the relative energy consumption can match the environmental impact.

At present, it is difficult to attribute specific renewable and energy-efficient resources to energy services and thus provide an impediment to new investments in such resources and technologies. The general assumption is that abundant resources and technologies are most cost-effective, even though they may not provide the relative energy service in comparison to the alternative and could cause further energy consumption. Abstaining from energy usage can be seen as the easiest and most cost-efficient way to achieve energy services.

Smart grid systems embody an important step in the process of energy management. They enable long-term sustainability by reducing energy usage and reliance on unsustainable resources. Smart grids employ the use of digital technology to improve reliability, security, and efficiency of the electricity system. This volume provides an in-depth understanding of how smart grid technologies can contribute to the only partial solution for combating climate change. A partial solution is to shift to an energy system using abundant and sustainable clean energy resources. A future electricity system will be a highly integrated energy system, which in turn provides a wide range of services to the customers. This will effectively enable the use of highly efficient end-use technologies and demand resources in delivering and optimizing energy services, which will, in turn, enhance the quality of life.

5.1 Demand Response

With an abundance of variable resources in the future energy system, two-way communication and automation at the customer end will be a valuable tool in integrating renewable energy and maintaining a reliable electricity system.

Advanced event notification provides a later option for turning off equipment, giving consumers plenty of warning and flexibility to turn off equipment at a specified time before the event. Technologies exist now that can enable a customer to program an energy management system to respond to electricity price changes or the availability of renewable power, but two-way communication between the utility and the consumer would allow the utility to automate changes in the customer's system based on the price or the amount of renewable power that is available. This can much more effectively shift load to off-peak times or to times when an abundant supply of renewable electricity is available, and away from times when fossil fuels are the dominant source of power.

Currently, demand response is implemented by direct load control using a one-way communication system between the electric utility and the consumer. The utility's control center sends a signal over a specified period, and the consumer's appliances or equipment are controlled by the utility at some time during that period.

Demand response programs will reduce GHG emissions by providing a price signal to the consumers of electricity, which will elicit a response that will change their electricity usage or move it to another time period. In addition to targeting greenhouse gas emissions, demand response can contribute to the reduction of peak load on the electricity system, which will reduce the need for expensive peaking generators and can, in turn, reduce the market price of electricity. Lower electricity prices will make it more competitive for the water pumping and space heating sectors to use renewable electricity as a clean energy substitute.

5.2 Advanced Metering Infrastructure

With AMI, utilities can measure consumption more frequently and, with the two-way communication, can implement time-based rates that provide consumers with economic signals. The data can be used to better understand energy usage, enhance the service to consumers, and increase the efficiency of the energy system. Static meters provide only total consumption and present available capacity whereas advanced meters provide information on when and how energy is used. This data can be useful in identifying the impact of certain technologies such as air conditioning, and assessing the potential benefits of energy efficiency or demand response programs. AMI can thus enable more informed and efficient use of energy.

Advanced Metering Infrastructure (AMI) provides utilities with a real-time tool to measure end-user energy consumption. AMI consists of advanced meters which are read remotely and two-way communication between the meter and the utility. An advanced meter is a meter that goes beyond the basic one-way measure of energy consumption to provide a benefit or service to the consumer or utility.

CHAPTER 6

Green Building
Technologies

An HVAC system is a crucial part of many buildings, particularly those in warm or cold climates where heating and cooling are necessary for occupant comfort. HVAC systems are also large consumers of energy, which is a barrier to greener buildings where high energy use is avoided. In North America, it has been estimated that heating and cooling account for up to 45% of home energy usage. It is likely that a similar figure is true for commercial buildings. The best method for reducing the energy impact of an HVAC system is to use less energy to heat and cool the building. Secondly, energy that is used should be sourced from a renewable source. This can be achieved through the use of solar panels to power the system. However, the most profound method is the reduction of energy use by designing a building that requires less heating and cooling.

The techniques for constructing a green building can vary widely, but one common differentiator of 'green' buildings from conventional buildings is the emphasis on energy-efficient systems that are sustainable to maintain. Green buildings use less energy, water, and natural resources, and create less waste in comparison to conventional buildings. This is accomplished through the implementation of many different strategies and techniques that fall under the title of Green Building

Technologies. This section of the essay looks at three of these strategies: an energy-efficient HVAC system (heating, ventilation, and air conditioning), passive solar design, and green roofs. Each of these strategies is used in many buildings globally and has proven to be an effective way of reducing the impact of a building on its environment. Each has various costing and effectiveness, but each shares the goal of achieving a sustainable building that has lower energy use and impact on the environment.

6.1 Energy-Efficient HVAC Systems

In the long term, it is highly conceivable that these technologies will become the standard for both residential and commercial heating and cooling, and it is noted that this is one of the most important areas for energy conservation in the building sector.

Yet while these technologies are promising, the most revolutionary potential lies in heat pump technology. This is because it can provide heating and cooling from the same piece of equipment, and is the most flexible in the way it utilises electricity and alternative energy sources. With this in mind, it is important to take note of recent developments in air source heat pump (ASHP) and ground source heat pump (GSHP) technology. Internet databases such as the Energy Star Program in the United States are now allowing consumers to compare different models of ASHP and central AC systems for energy efficiency. ASHP systems in Europe have shown potential to cut electricity use for heating by more than half when compared to electric resistance heating, and GSHP systems have been shown in several studies to provide heating and cooling using 25-50% less energy than high efficiency gas systems and over 30% less energy than electric resistance heating and standard air conditioning.

New technologies that drastically reduce the energy consumed by HVAC systems are likely the most cost-effective way to save money, and reduce a building's energy consumption and associated emissions. In Japan, the market for energy efficient heating systems is already very competitive, with a number of energy efficient space-heating technologies on offer. One such technology uses a pump to move heat into a house, rather than combustion, and can deliver roughly 3 times more heat energy than the electrical energy it consumes. Another system known as "Gas Absorption Air Conditioning" can provide space heating and cooling, as well as hot water. Due to the fact that it transfers heat to provide cooling, and can utilise waste heat from other sources, it can deliver 1.5-2 times more cooling and heating energy than the energy it consumes. This is equivalent to air conditioners with a COP of 3-4, yet the absorption system does not use combustion, and thus does not increase carbon emissions in the summer, and contributes to a decrease in emissions in the winter if the electricity/gas it uses is from a low carbon source.

While some may argue that the heating, ventilation, and air conditioning (HVAC) system has relatively little to do with a building's energy consumption, others beg to differ. Research from the American Council for an Energy-Efficient Economy revealed that home heating accounts for roughly 42% of residential energy consumption in the United States, making it the largest energy expense in any home. A testament to this, it is projected that the world will install 700 million air conditioners by 2030, and 1.6 billion by 2050. To keep the greenhouse emissions associated with these numbers in check, and the energy costs to a minimum, it is clear that a revolution must take place in the way that buildings are heated and cooled.

6.2 Passive Solar Design

Isolated gain systems store the sun's heat in a location other than the living space. An example would be a sun space. This is not suitable for use in housing.

Indirect gain does not pass through a window but is absorbed by a wall and reradiated on the other side to be absorbed by the living space. This method is less effective in comparison to direct gain.

Direct gain is the heat from the sun admitted through the windows. It is absorbed by the home's mass (such as walls and floor) and then released at night when the heat is most needed. This is essentially how direct gain is effective: by using the sun's heat stored in a thermal mass for delayed heat release.

Passive solar design refers to the use of the sun's energy for the heating and cooling of living spaces by the use of direct and indirect gain. Solar radiation travels through a glazed window and is absorbed by a mass, which then reradiates it at a longer wavelength. This radiation cannot pass back through the window, thus heating the living space.

Passive solar technologies are simple, have low risks, and moderate costs. However, the overall efficiency needs improvement due to the unpredictable nature of when energy is needed and when it is available. Many passive solar building design features were contrived in the 1970s, though there has been little development since that time. Many are now considering revisiting these concepts in an improved and modified way, particularly with the use of modern computational tools.

6.3 Green Roofs

The concept of green roofs is not a new one; however, it remains an idea that, up until recently, has been overlooked. The study and practice of green roofs in Germany and other European countries have demonstrated that green roofs can reduce the need for air conditioning in summer and also reduce heat losses in winter. This, in turn, leads to a reduction in carbon emissions and a direct impact on stemming climate change. This is all a result of the insulating effect of soil and vegetation. According to researchers in the USA, extensive green roofs can reduce a building's energy consumption by up to 6%. A Canadian study, on the other hand, found that green roofs, cool roofs (roofs that reflect more sunlight and absorb less heat than traditional roofs), and hybrid roofs (those that contain elements of both green and cool roofs) could reduce energy use for air conditioning in the greater Toronto area by over 50%. The three roof types used in this study were found to have different potential in terms of temperature reductions and cost. However, all were found to be more effective than conventional roofs, with the green roof providing the best insulation. This is good news for our future. With more buildings being built and less green space, cities are becoming heat islands that have a significant impact on the warming of our planet. In fact, air conditioning, the energy source that will be saved from the use of green roofs, is ironically a large contributor to the heat island effect. It is a vicious cycle where the cooling of indoor spaces leads to the warming of outdoor spaces. It is hoped that the use of green roofs becomes widespread and a requirement for all new, large building developments. This would lead to a significant reduction in energy consumption and therefore a reduction in greenhouse gas emissions.

A green roof, also known as a living roof, eco-roof, or planted roof, is a roof that is partially or completely covered with vegetation and a growing medium, planted over a waterproof membrane. It may also include additional layers such as a root barrier and drainage and irrigation systems. We will use this definition of green roofs when referring to them throughout the text.

CHAPTER 7

Waste Management Solutions

Recycling technology is developing to enable more sophisticated materials to be recycled. For example, developments in molecular separation techniques allow the almost complete separation and re-manufacture of plastic polymers. Technologies such as these, which are economically viable, are the most sustainable way of using materials. Smarter product design can assist this. For example, using a single material in product design rather than complex assembly of many components, thus enabling easier disassembly and recycling. The concept of "Design for Recycling" is thus a key goal for a sustainable future and aims to close the loop on as many materials as possible.

Recycling is perhaps the most common technique people are aware of. Whilst the benefits of diverting waste from landfill are clear, there are only a small number of materials like paper, wood, glass, and metal which can be recycled in a closed loop back into a similar product. Materials like food packaging and plastic bags can really only be recycled into a lower quality product (e.g. plastics into park benches and building products), and the process is often "downcycling", as most recycled products will still be more costly than using virgin materials which are currently abundant due to relatively low prices of production and consumption.

7.1 Recycling Technologies

Another new and emerging technology in recycling is biorecycling. Biorecycling studies the biological value of waste and tries to reintroduce it to our biosphere. Organic waste is the main target of biorecycling and it is defined as waste which comes from the earth and is readily biodegradable, i.e. paper, cardboard, food, and green waste. These materials are usually sorted and transferred to a landfill which is not good practice, especially for green waste. An alternative method is composting, which takes green waste and other organic waste and converts it into a soil conditioner. Though composting is useful, a more efficient method lies in enzymatic processes and fungal cultures. Enzymatic processes are high in specificity and low in waste production. An example is cellulose hydrolysis in which cellulose from material is broken down into glucose. Fungal cultures are used for fermentation and are useful in creating many products. An example is the production of ethanol from paper waste. This high-grade form of recycling for organic waste can reduce the amount of waste sent to landfills, decrease pollution, and conserve energy.

Even though the practice of recycling is good, statistics show that the amount of waste deposited in landfills has not decreased. This is because the processing and yields from the recycling plant are not efficient enough. The reason for this may be due to energy costs and production costs of the recycled material. One possible solution to this is nanotechnology, and more specifically, nano-filtration. Nano-filtration membranes have been developed to separate metal ions and organic matter from water. If this technology was used with the same purpose in mind, to separate the waste into basic components, it could reduce the amount of waste and produce a higher quality of recyclable material. By doing so, the costs of processing and recycling would decrease, and ultimately reduce the amount of waste deposited in landfills. This is because higher quality recyclable material equates to higher priced material and less material being rejected, so the economics would drive companies to recycle. However, this method may increase the waste of the nano-filtration membranes due to fouling and corrosion.

So a method to combat this would be required and it is suggested that the use of magnetic nanoparticles to aid the cleaning process is the way forward.

The reward of technological development is that it brings better techniques to help us achieve our goals. Our current goals are set on sustainable futures and the technologies developed in this field are certainly innovative and promising. The field of waste management is being influenced greatly by technological advancements. With the promise of recycling being a widely accepted and a global practice, it is important to understand the recycling technologies that are being developed and how they work.

7.2 Waste-to-Energy Conversion

Waste-to-energy (WtE) involves the generation of energy in the forms of heat or electricity from the primary treatment of waste. Global interests in WtE have been increasing in recent years as an alternative to the disposal of waste in landfills, and has also gained significant attention throughout Europe with the implementation of the Waste Incineration Directive in 2000. Various WtE technologies have been developed in response to the negative implications of landfill and the benefits of energy generation. These technologies include incineration, pyrolysis, gasification, and other emerging thermal and non-thermal chemical treatment techniques. The diversity in technological approaches reflects the complexity of waste as a potential energy resource and the need to tailor technology to specific waste streams.

Global society, and developed nations in particular, have shown an increased trend toward waste generation, with the general assumption that disposal of wastes is not the responsibility of the waste generator. A drastic increase in resource consumption and a throw-away society has led to the creation of many different kinds of wastes, some hazardous and some not, with the general trend of disposal being in landfills. Although waste disposal is the current strategy employed, it is not sustainable and does not represent best practice, demonstrated by the negative impacts on the environment and communities that host waste management facilities.

CHAPTER 8

Sustainable Agriculture Practices

Precision agriculture is based on observing, measuring, and responding to inter and intra-field variability in crops. The goal is to apply the right treatment in the right place, at the right time. Often the input costs and the potential environmental damage are much reduced. This can be contrasting to organic farming, depending on how resources are considered in the latter. In essence, precision farming is about providing farmers with more control, finesse, and options for decision making for the management of their crops. It is not necessarily about using technology or machinery, but is a systemic approach to farming. The essence of precision farming is built around tools to assist decision-making for the farmer, leading to increased quantification of effects and more accurately controlled experimentation. In turn, this emphasizes the importance of monitoring for the successful implementation of precision farming, where the results of measuring and decision-making are closely related.

Organic farming has the surround sound of a good thing. It is a method of farming that works at ground level (i.e., beneath the ground) in ways that are holistic and ecologically sound. The fundamental concept of organic farming is to work with the environment rather than against it. It is a style of farming that avoids the use of manufactured fertilizers, pesticides, growth regulators, and additives. Organic farmers rely on developing a healthy, fertile soil and growing a mixture of crops. An important characteristic of organic farming is the high degree to which farming inputs are matched with production outputs. The premise being that if a farm is truly sustainable, the resources used by the farming operation should be replaceable by the farm itself, thus having little or no effect on the external environment. In keeping with the holistic aspects of organic farming, it is important to look at the large field in which farming occurs. Here, there are contrasting views as to the extent of any positive or negative effect that organic farming methods have on the environment. This leads to the study in comparison to conventional farming practices as to its overall sustainability.

8.1 Organic Farming

This definition from the FAO encompasses the theme of sustainability within this essay. Organic agriculture methods offer the best potential and most practical means of realising genuine sustainability in agriculture. Whether in the periphery or in core debates, sustainability is in embracing and efficient use of resources to meet human needs while ensuring the long-term health of the natural environment. Sustainability is in the comparison of the efficacy of solutions to problems. Sustainability is in the economic viability of the methods. Organic agriculture leads to sustainability in all of these areas and for a wide range of products. It is less damaging to the environment, more tightly cycling resources and relying on renewable resources, and has generally provided farmers with a more secure income. The sustainability of organic farming is able to meet the needs of the present without compromising the ability of future generations to meet their needs. This is the goal of sustainable development, which has been labelled as the pathway to sustainability for farmers.

The Food and Agriculture Organisation (FAO) of the UN defines organic agriculture as "a holistic production management system which promotes and enhances agro-ecosystem health, including biodiversity, biological cycles and soil biological activity." It emphasises the use of on-farm cultural, biological, and mechanical methods, as opposed to synthetic inputs. The primary goal of organic farming is to optimise health and productivity of interdependent communities of soil life, plants, animals and people. It upholds the precautionary principle and seeks to support the long-term health and sovereignty of humans and their environment.

8.2 Precision Agriculture

For example, avoiding over-application of nutrients can prevent run-off into water systems and avoid the release of greenhouse gases (methane and nitrous oxide) that is associated with excessive fertilizer use. Precision agriculture takes many forms and is suited to different scales and types of farming. A number of these practices have the potential to save natural resources and reduce the impact on the environment.

In contrast, PA aims to measure and apply the right amount of inputs at the right place and the right time, using a combination of on-farm information and technology such as GPS, remote sensing, and GIS. By acting on accurate information, it is possible to avoid waste through over-application or under-application of farm inputs. This can lead to increased resource use efficiency – with potentially less impact on the environment.

Precision agriculture (PA) is one of the many modern farming practices that seeks to implement the sustainability and environmental stewardship that is one of the key goals for the future of agriculture. The term precise is meant to differentiate this approach from traditional farming practices, where farmers often apply whole fields with much more input (seed, water, fertilizer, pesticides) than is required. This is more of a blanket approach and increases the risk of wasting resources and money.

CHAPTER 9

Water Conservation Technologies

Drip irrigation is a type of micro-irrigation that has the potential to save lots of water and nutrients by allowing water to drip slowly to the roots of plants, either onto the soil surface or directly onto the root zone, through a network of valves, pipes, tubing, and emitters. Although it may be more costly to install a drip irrigation system, it has the potential to save significant amounts of water. This has shown to often lead to a higher quality crop, the reason being that there is less water to damage the fruits and a constant delivery of water to the roots. Additionally, it has been shown that weed growth is less in a drip system compared to other methods of irrigation. This is because there is less water in the rows, the area between the plants has less moisture, and therefore the seeds of weeds do not sprout.

Rainwater harvesting is the accumulating and storing of rainwater for reuse before it reaches the aquifer. It has been used to provide drinking water, water for livestock, water for irrigation, as well as other typical uses. Rainwater collected from the roofs of houses and local institutions can make an important contribution to the availability of drinking water. It can supplement the subsoil water level and increase urban vegetation. Water collected in the tanks is often used for domestic and garden purposes, freeing up the main water supply for economic purposes (such as pumping to a higher altitude) and higher quality uses. Using rainwater inside the house for drinking, cooking, and bathing could improve health. In some regions, it is required that these tanks be open to permit inspection by health authorities.

9.1 Rainwater Harvesting

Harvesting rainwater offers many benefits. In addition to providing a safe source of water and reducing the strain on the municipal water supply, it is an economical way of obtaining or supplementing the supply of water for the end user. Taxes are seldom levied on the collection of rainwater, as it is mostly used as a method of controlling stormwater runoff or to decrease an area's flood risk, though this can change depending on the area. This can lead to higher costs for the end user, but when compared to the cost of rainwater harvesting to the cost of a utility company, well water, or water from an independent system, it is clear that rainwater will likely cost less.

Rainwater harvesting is being used throughout the world for many different purposes, such as landscape irrigation, gardening, and lawn maintenance, and as the primary water supply for entire households.

Rainwater harvesting is a technique widely used in the world for collecting and storing rainwater to use for later. This method can be as simple as collecting rainwater in a tank and using it for watering plants or for washing, or it can involve storing rainwater underground. One of the best ways to harvest rainwater is to build a system of gutters and storage tanks to store the rainwater to be used when it is needed. This can be a simple system just to store a little water for a drier time, or a complex system to store large quantities of water to be used all year round.

9.2 Drip Irrigation Systems

Drip irrigation is a method of applying moisture directly to the root zone of the plant, a few drops at a time. The goal of using this method is to apply water at a rate that is less than the infiltration rate of the soil to avoid surface runoff and water loss below the root zone due to deep percolation. This method allows water to be applied uniformly and slowly, making it possible to keep the soil moisture relatively constant over a period of time. This technology has proven to be efficient in water usage. Drip irrigation is 95% efficient, in contrast to sprinkler systems which are 80% efficient, and surface irrigation systems which are 50% efficient. The efficiency of the irrigation method, combined with the ability of the plants to take in the right amount of water they need, will result in substantial water savings for the farmer. The recent implementation of using a computerized system to monitor the soil moisture and automatically turn the system on and off has further improved its efficiency. In line with the understanding that soil can absorb a limited amount of moisture per unit time, this method can regulate soil moisture within an ideal range for the plant, between the field capacity and the permanent wilting point of the soil. The result is that the crop transpiration will be more water-use efficient, and there is no leaching of water below the root zone.

CHAPTER 10

Circular Economy Approaches

10.2 Material Recycling In comparison to the linear economy model, more of the end-of-life materials get one more chance at the market before being consigned to a waste disposal system. Material recycling allows for waste materials from production and consumption to be transformed into new useful products, or regenerated materials are re-entering the production process. This contrasts with the traditional recycling methods that involve simple down-cycling or the extraction of materials into a lower quality product still receiving the same fate. Two types of recycling are materials and energy recycling. Energy recycling focuses on energy recovery from materials and uses processes such as incineration. However, this is not an ideal method in a circular economy as material is destroyed and opportunities to maximize recycling are lost.

10.1 Product Life Extension The idea behind a product life extension is to increase the length of time a product remains useful. Consumers' culture should change from a throwaway society to one where products are maintained or upgraded. Companies should produce in a modular design, allowing for more efficient upgrading or repairing by the consumer. An example of a product that has increased length of use due to modular design is in mobile phones. Customers are more frequently changing software, and their phones are capable of tasks that models 5 years ago never could. Some mobile phones have parts that can often be replaced, allowing the same shell of a phone to be used for a number of years. This is compared to a non-modular design phone, which ends up being discarded quicker due to parts becoming unavailable.

In a circular economy, everything in the cradle-to-cradle loop should run in a closed economic system. The value of products and materials is maintained for as long as possible, and the generation of waste is minimized.

10.1 Product Life Extension

There are several strategies under PLE, such as maintenance, repair, remanufacturing, and remarketing. Technologies that can enable PLE include durable product design, enabling technologies, product management, and resource management. There are already a number of well-established PLE systems globally, including Xerox and Caterpillar. Xerox has its own product line entitled "Green World Alliance," which sees printers, photocopiers, and multifunction devices remanufactured instead of being recycled. This has resulted in 24,000 units being remanufactured, saving an estimated 7,800 tonnes of waste from landfill. Caterpillar is mostly known for remanufacturing heavy vehicle engines. The practice has resulted in a 70% recycling rate of engine cores. This requires less raw materials and uses less energy in the manufacturing process, which is a win for both the company and the environment.

Product life extension (PLE) is an essential aspect of a circular economy (CE) and involves the reuse of a product to lengthen its life for multiple use cycles. Utilizing the practices of PLE can prevent waste, keep products at their highest value, and utilize the embedded energy and materials. This, in turn, can lead to a reduction in the need for producing new goods, which often require higher quantities of resources, energy, and materials.

10.2 Material Recycling

The concept of self-sorting smart materials has been proposed, in which identification and separation of waste materials can take place with little or no human intervention. This can be achieved through a variety of technologies, including embedded identification tags, the use of electromagnetic properties, spectroscopic analysis, and the use of magnetic and eddy current separation techniques. Automated sorting systems can greatly reduce manufacturing costs for recycled products and ensure a continuous feed of high-quality materials into reprocessing plants. Further research is required to make such systems economically viable and to investigate the life cycle environmental impacts of various self-sorting material and product scenarios. An EU-funded project titled SMaRT-EU has performed research into the development of self-sorting smart labels for packaging that can undergo automatic detection and separation using a combination of identification techniques.

Each year, society generates millions of tonnes of waste with increased levels of material consumption and distribution. This is largely manifested through packaging, disposable items, home electronics, and large-scale equipment. Waste collectors are provided with bins for mixed recycling; therefore, there is very little incentive to separate different types of waste at the source. This mixed recycling is often of poor quality, and recycled products can end up simply being thrown into landfill due to contamination. With concerns over raw material availability and price increasing, future scenarios will likely necessitate the collection, sorting, and recycling of specific materials to minimize energy costs in manufacturing secondary products. Material recycling is a key process in the creation of closed loops for sustainable materials and products. By extracting and reprocessing materials into new products and packaging, the value of the material is extended, energy and resource consumption is reduced, and environmental impacts are minimized.

CHAPTER 11

Climate Change Mitigation Strategies

Afforestation is the practice of planting trees on land that is not a forest, or has not been a forest for a long period of time. This method increases the number of trees on the planet and therefore increases the carbon sink. When a tree grows it removes CO_2 from the atmosphere and uses it to form its structure, this carbon remains stored in the wood even after the tree has reached the end of its life. The carbon will only be released back into the atmosphere if the tree is burned or left to decompose. Therefore increasing the number of trees and preserving existing forests will in turn remove more CO_2 from the atmosphere and store it for a long period of time. They go on to say the cost for this is between $10-50 per tonne of Carbon, which is competitive with other mitigation options.

The 11th section of the essay examines a few of the solutions available today to mitigate climate change. This is a controversial issue because many people believe it is better to just stop burning fossil fuels instead of using another method to mitigate the effects. One of the most promising methods of mitigating climate change is using Carbon Capture and Storage (CCS). The idea behind this is to capture the carbon dioxide emitted into the atmosphere before it has a chance to disperse, then compress it and transport it to a location where it can be stored, so it will not enter the atmosphere. The captured carbon is usually compressed and transported to an underground storage site. This directly reduces the amount of CO_2 in the atmosphere and has the added benefit of extending the life of certain depletable resources. The Intergovernmental Panel on Climate Change (IPCC) has stated "large scale CCS deployment should be a key mitigation option for 2030 and beyond in scenarios aiming to stabilise atmospheric concentrations..." This is quite a bold statement considering it is still an emerging technology.

11.1 Carbon Capture and Storage

Carbon capture and storage (CCS) is a method that is currently under development, with the intention of mitigating the effects of climate change. The process involves capturing CO_2 emissions from fossil fuel power generation and other major point sources such as cement production, and then storing it deep underground to prevent it from entering the atmosphere. It is believed that this technology will be essential in achieving a globally low carbon economy. There are three separate ways in which CO_2 can be captured: post-combustion, pre-combustion, and oxy-fuelling. Capture technologies have been developed by both the oil industry and natural gas industry. The CO_2 is then transported in its liquid form via road tanker or pipeline to a storage site. Once at the storage site, the CO_2 is pumped into dedicated underground storage formations where it will be stored securely for a vast period of time. Storage locations are varied; saline aquifers are the most common storage location and have been proven to store CO_2 effectively. Oil and gas fields are another potential storage location, using the same technology that has been used to extract the oil and gas. The stored CO_2 can stop any further emissions and help to increase the amount of oil and gas extracted. Another storage location that is less common, but has great potential, are coal seams, where the CO_2 can be stored through enhanced coal bed methane recovery.

11.2 Afforestation

Afforestation is a means of mitigating the enhancement of greenhouse gases in the environment. It is the same as planting trees, which will absorb the CO2 from the atmosphere, thereby reducing the amount of CO2. It is a cost-effective and simple way to remove CO2 from the atmosphere. It is to create a forest in an area, which thereby affects the microclimate of the area, making it last longer than before. It is the process of planting trees in an area that may have never been a forest. For example, in an empty tundra area, the plantation of conifer trees can result in increased absorption of sunlight. But specifically, it is an area of temperate grassland with infrequent land use that can endure for a long time. However, the increased radiation can result in drying and eventual changes in the trees. These changes, irreversible, give the idea of afforestation. Like these changes, they are not necessary. The idea of a more recent afforestation project shall be in Brazil to stop the depletion of the Amazon forests. This process will mimic, in every way, the climate change that occurs slowly around the equator and tropical areas. After completing thorough impact assessment studies – involving soil, water, and biodiversity – a typical forest plantation in Malaysia will receive a forest management license for a concession period of 60 years. Most countries are signatories to one or more international treaties and protocols aimed at biodiversity conservation and sustainable forest management. This is a key criterion for wood products exported to the European Union. The project will plant fast-growing mixed dipterocarp species, which will be harvested later for timber and pulp industries. At the end of the concession period, the forest will be converted to high conservation value forest or handed back to the Malaysian government. The fear, of course, is the long-term commitment and ability to change according to global and national politics.

CHAPTER 12

Sustainable Manufacturing Processes

An example of this would be the computer technology used in car manufacturing at the Ford Motor Company. This technology is used to plan every step of the manufacturing process and thus optimize it to be as efficient as possible. This will result in a manufacturing process using minimum resources and time, and will also find ways to correct mistakes in an efficient manner without wasted product.

The level of innovative technology used in lean production lines complies well with the idea of sustainable manufacturing. This is due to technology being used to find ways to produce goods in a shorter amount of time using fewer resources, in contrast to mass production lines which aim to produce as many goods as quickly as possible. This will inevitably use technology to catch up with extra demand in the case of an error in production. Too many resources used in the production of a good to be wasted would result in the waste of the product being made. This compares to the efficiency of lean production, where any surplus product goes on sale at a later date.

Lean production lines often produce less wasted product than mass production, as there is a lower margin for error in producing extra product to meet unforeseen demand. In the instance where over-production does occur, goods are saved to be sold at a later date and do not go to waste. This reduces the amount of resources used for the production of goods.

Sustainable manufacturing has grown in the consciousness of companies as more companies realize the need for sustainability. Lean manufacturing is a process whereby resources, particularly human labor, are used as efficiently as possible. This contrasts with traditional production methodology, whereby if a good sells, more will be produced. If a good doesn't sell, the production line will be shut down, resulting in idle workers. This allows lean methodology to be better for the environment compared to traditional methods.

12.1 Lean Manufacturing

Lean manufacturing is a set of production methods and tools to improve the efficiency and effectiveness of a company's operations by eliminating waste and improving quality. The ultimate goal of lean manufacturing is to increase the value of products or services for customers. This is done by considering what the customer is willing to pay for and identifying processes and resources that add no value from the customer's perspective (i.e. are wasteful). A product or service consists of both value-added and non-value-added activities. This will often involve a rigorous analysis of the current state, designing a new improved future state, and then implementing a solution to move from one state to the other. Lean thinking can be implemented in a number of environments ranging from production to service and public sectors. The key to the successful implementation of lean manufacturing is a long-term commitment and determination from top management at all levels. Lean manufacturing has five key principles. Specify value from the standpoint of the end customer by product family. Identify all the steps in the value stream for each product family and eliminating steps that do not create value.

12.2 Green Chemistry

The concept of green chemistry At its essence, green chemistry is the practice of minimizing environmental damage as a result of chemical processes. This is achieved through the design and adaptation of new environmentally benign products and processes. The twelve principles of green chemistry include the prevention of waste, the design of less hazardous chemical processes and products, the use of safer raw materials, and the ultimate aim of creating completely sustainable chemical processes. Dr. Mike Pitts, who currently serves as the Executive Director for the Clean Technology Centre in Wales and has an extensive background in the field of green chemistry, has stated that there are many reasons that companies should invest in green chemistry. As well as the widely acknowledged environmental gains, these reasons can include better use of resources in order to drive benefits directly to the bottom line, increasing market share by defending market access or opening new markets, and achieving compliance wins. This often comes as music to the ears of environmental chemistry researchers, who will understandably struggle to gain industry involvement in their research if they are not able to provide the potential for industry-driven self-sustenance.

CHAPTER 13

Environmental Monitoring and Analysis

Remote sensing is a method of data collection that is carried out using aerial or satellite photography. Relatively inexpensive, remote sensing provides a means of collecting data for large areas at regular intervals. The satellite and aerial images can be used to track changes in land use and land cover over time. In the context of biodiversity, remote sensing can also provide an indirect means of assessing and monitoring levels of genetic variation. This is achieved through an understanding that areas of high genetic variation are often correlated with higher species diversity and the co-evolution of communities. Specific habitat types can also be isolated on a regional basis, with a view to assessing the conservation status of the area and ultimately identifying the type and extent of human impact upon an area. Remote sensing has been identified as a valuable tool in the monitoring and control of invasive alien species and can be an effective way of identifying success following management initiatives. An example can be seen from remote sensing mapping of Australia to reveal the current national distribution of the introduced European rabbit (Oryctolagus cuniculus), which high-lighted areas of partial success and ultimately led to the identification of the areas where the biological control agent myxoma should be targeted. Measurements from remote sensing can also be continually integrated with other data types to monitor change, assess and predict future scenarios, and ultimately provide a basis for informed decision making in the field of conservation.

13.1 Remote Sensing

For some projects, it is possible to use already available data, which is often free. An example is the use of LANDSAT data that can be downloaded from the USGS website.

An example of its use is in detecting vegetation health and distribution. This has been done in Canada to model the carbon cycle and to identify how much CO_2 is being absorbed by the vegetation, an essential factor in climate change studies. Through the use of data such as this, we are able to better analyze specific interactions within the environment and form a better understanding.

Remote sensing allows scientists to access various data stored, which can be used to create models and simulations. This kind of technology has widespread application as it is cost-effective and can provide a wide range of information for analysis on a large and small scale. It also has great potential for monitoring and identifying changes in the environment. The types of data that are accessible using remote sensing are vast, ranging from elevation and temperature to plant health and land use.

Remote sensing was originally developed during the 1960s for military use, and the technology has since proven itself to be a cost-effective means of providing reliable and timely information for decision-making. Remote sensing, also known as satellite imaging, works by scanning the Earth with sensors that make use of specific wavelengths. A sensor collects data by detecting the energy that is reflected from the Earth. The data are then transmitted to a ground station, upon which they are sent to off-site servers where they can be accessed by the public.

13.2 Air Quality Monitoring

As we are expending a lot of concern focusing on the health effects caused by very low air pollutant levels, it is necessary to have air quality monitoring tools that are highly sensitive and have a large detection capability. This is because most health effects from air pollution are chronic and occur from long term exposure at low to moderate pollutant levels. High acute pollutant exposures that lead to adverse health affects are usually due to sudden environmental occurrences such as forest fires or industrial chemical spills. Step one for monitoring air quality would be to know what specific air pollutants are going to be monitored and the main reasons for monitoring these. Next are the considerations of location and duration of monitoring. Finally choosing the right tools for the job and interpreting the data that they provide will all affect how worthwhile the monitoring project is in terms of reliability and information that can be attained.

Nowadays there are many types of air quality monitors offering a wide array of features and functions. However, a lot of the lower cost monitors are not worth purchasing due to their lack of accuracy, reliability, and limited range of compounds that they are able to detect. Some of the common types of air monitoring systems include ones that measure the concentration of a specific air pollutant (example SO_2, NO_2), monitors that measure and record toxic air pollutants in the ambient air (usually these are specified for indoor air quality) and complete air assessment tools which take into account meteorological and local geographic influences on air quality.

Indoors and outdoors, the quality of air has tremendous influence on our health, safety, and comfort. Air pollution can cause myriad negative health effects, including respiratory problems, exacerbation of current health problems, and headaches. So it is important to have air quality monitoring that is reliable, comprehensive, and accurate over time. In many cases, detection of high pollutant levels can prompt remedial actions, protecting human health and the environment.

CHAPTER 14

Policy and Regulatory Frameworks

A feed-in tariff is a policy mechanism that offers a guarantee of payments to renewable energy producers. Payments can be made on either a per kilowatt-hour basis or through a contract for the energy produced. This mechanism has been successful in the wind industry in Germany. These policy mechanisms are an effort to create an incentive for the use of renewable energy, increase energy security, reduce environmental damage, and lower the costs of renewable energy. An in-depth explanation of policy mechanisms can be found in "Climate Change 2007: Mitigation of Climate Change" Chapter 13.

Net metering is a policy which is supported in 30 US states that allows those who generate their own electricity from renewable energy sources to sell excess electricity back to the grid at retail rates. In 2008, the state of Hawai'i launched the most ambitious consumer renewable energy initiative to date.

Some of the common policy and regulatory frameworks that have been established around the world to increase the use of renewable energy include renewable energy obligations, financial incentives, net metering, and feed-in tariffs. Renewable energy obligations are regulations that require increased production of energy from renewable sources and often include a statement on the relative amount of energy that must be met through renewable energy. Financial incentives can be used to encourage private investment in the development of renewable energy technologies. Examples of financial incentives include rebates, tax credits, and production payments.

14.1 Renewable Energy Incentives

The United States and countries with a liberalized energy market have tended to favor market-based incentives such as Renewable Energy Certificates (RECs) and competitive grant programs. The advantage of RECs is that it creates a sustained income stream for renewable energy projects by requiring electricity retailers to procure a certain percentage of their energy from renewable sources or purchase RECs in lieu of this. While this has been effective in driving renewable energy investment in some states, it can be vulnerable to changes in legislation and not always adequate in size to provide the necessary incentive for large-scale investment in renewable energy.

The two principal reasons why governments provide incentives for the deployment of renewable energy are to reduce greenhouse gas emissions and to increase the security of their energy supplies. These reasons are shared by the public and environmental groups and drive the political process behind renewable energy incentives. The way in which these incentives are provided depends very much on the type of political system in place and the influence of different stakeholders on the policy-making process. The survival and success of the renewable energy industry is contingent on its ability to navigate through the complexities of political processes to ensure that effective incentive programs are implemented and do not fall victim to changes in governments or economic downturns.

14.2 Carbon Pricing Mechanisms

In order to establish which of these views is well-founded, it is important to understand the relative effectiveness and efficiency of separate incentives and that of a carbon price. A highly detailed comparison of which is beyond the scope of this report.

While this argument has merit, it must be remembered that a carbon price is essentially a broad incentive spanning all technologies and all sectors. Though it is important for governments to supplement a carbon price with other instruments that fast track specific technologies which are particularly far from being competitive with the incumbent method, it remains inefficient to have a multitude of separate incentives all with differing administration methods, eligibility criteria, and rates of abatement subsidizing different technologies in different countries.

In order to stimulate technological progress and extract maximum innovations targeted at climate change, a wide variety of analysts and researchers advocate the use of diverse, highly focused incentives as opposed to setting a carbon price and leaving it up to the market to determine where and how abatement is achieved. This view contends that a price is not enough as there are so many different technologies across so many sectors that are at such different stages of development. Thus, certain technologies may not be given the research and investment required to bring their costs down closer to conventional alternatives without targeted incentives.

CHAPTER 15

Education and Awareness Programs

Numerous studies have shown that raising awareness about environmental issues and educational programs significantly influence public perception and attitudes about the environment. These findings suggest that education may be the key to affecting long-term behavioral changes. For instance, one study found that over the past two decades, environmental education has led to an increased level of concern over the environment. This increase in concern may ultimately promote environmentally friendly behavior and influence public policy. Another study asserts that there is a distinct correlation between environmental knowledge and pro-environmental behavior. However, this study confirmed that the most significant influence on behavior is not simply knowledge itself but rather knowledge that is accompanied by a strengthening of personal environmental ethics. This indicates that programs aimed at promoting environmental awareness and encouraging an ethic of stewardship may be most effective at eliciting behavior change.

15.1 Green Technology Training

At the highest level of education, research universities are investigating and developing the cutting edge of new technologies for sustainability. This includes new applications of old ideas such as materials science research into natural composites, as well as the development of entirely new technologies being examined at the Institute for Resilient Infrastructure at the University of Colorado.

A variety of post-graduate programs offer engineers and architects the opportunity to specialize in sustainable design and building. Many of these are certificate programs consisting of a series of short courses. Examples include the building science and sustainable design certificate program at the University of Minnesota and the passive solar design and construction certificate at Colorado State University. Other programs offer a master's degree in a field of sustainable technology. This is the case for the energy-efficient building program at Stanford University.

Instruction in green technology is taking place at many levels and in many places. At the community college level, there are programs that range from a course or two in sustainable design for architectural students to two-year technician training in building science. Four-year programs in engineering with a special emphasis on energy conversion and materials science are being offered at an increasing number of universities. Such programs provide the basic education that one would receive in a traditional engineering program, but the skills are imparted in the context of materials and methods that are compatible with a sustainable future.

15.2 Public Outreach Campaigns

Education is an area which often receives inadequate attention in the effort to further green technology. Public education is an excellent tool for increasing awareness, but the absence of an organized effort to inform the populous about recent advances in greener technology results in public knowledge that is often dated and riddled with misconceptions. Public service announcements or educational programs are a fundamentally simple method of informing the public about changes in technology, but it is a method which is rarely used to promote green technology. In the absence of broad public knowledge, available green technologies often go unused because the average consumer is unaware of the benefits. It is easier for an individual to decide to take action if he or she is well informed on a topic. Public knowledge of environmental issues often leads to a sense of helplessness because the problems seem too large for any individual to solve. But equipping the public with information on new technologies that they can utilize in their everyday lives to make a difference might offset this pessimism and result in an increased consumer demand for green products. A very effective way to generate this demand is by targeting the newest generation of consumers. It is often the younger demographic that is more aware of current events and informed about issues like the environment. Creating a generation of consumers who are aware of green alternatives and are mindful of the environmental impact of their decisions will generate a lasting consumer demand for green products. This group is also the most impressionable, and getting them started on the right track in considering the environment when making decisions can create a lifelong habit.

CHAPTER 16

Conclusion

This look at several technological solutions available to help secure a sustainable future has shown that there is not one catch-all solution, but a complex selection of ideas with diverse relative strengths. Certain ideas may be more appropriate than others with respect to specific societies or global regions, and certain ideas may be essential precursors to others. It will take sustained initiative and open-mindedness, combined with well-informed selection and implementation of technologies, to pave the way to a more sustainable future. In more optimistic terms, it is possible to say that appropriate technological solutions for sustainable development actually do exist, and may merely be awaiting a setting of sufficient global will and determination to see them through into reality. Success in doing so is imperative if humans wish to meet their needs in the present without compromising the ability of future generations to meet their own needs.

The conclusion suggests that there is cause for optimism in the quest for a sustainable future, due to the abundance of technological solutions already available or on the verge of being implemented. The wide variety of technologies and approaches mentioned in this essay are proof that all is not yet lost, and that there is still time to change the way society is impacting on the environment. It is generally agreed that humans will not stop setting out to meet their needs and desires, so the only alternative is to develop a means of doing so sustainably. This does not mean reverting back to a pre-industrial state of technology, but using our technological prowess to develop more efficiently and less environmentally damaging alternatives. The technologies outlined in the paper are proof that this is a realistic goal. High profile examples such as hydrogen-powered vehicles, and long-term power storage in the form of synthetic hydrocarbon fuels, can serve as beacons of hope and initiative to draw people towards a greater appreciation for sustainability in technological development. The success of these technologies in both the private sector and public perception may create powerful snowball effects towards wholesale change in industry paradigms. Step by step climate engineering may be viewed as controversial or defeatist in terms of environmental preservation, however it may prove essential in rectifying past mistakes and potentially buying time in more ambitious initiatives to halt or reverse climate change. The example of re-forestation as a means of carbon sequestration shows that humans can still learn from nature, and how to rectify imbalances caused by earlier mistakes. In contrast, solar radiation management shows how humans may take on an altering role of managing nature to preserve desirable conditions for life. The fact that there is such a high density of ideas and proposed solutions suggests that it is not too late to change our course, though decision making and initiative will be core to their success.

Ways to a Sustainable Future

9 798330 617449